Influence of pets, rural/urban environment and dairy intake on oral microbiome

Mario Rodríguez Peña

Title: Influence of pets, rural/urban environment and dairy intake on oral microbiome

Author: Mario Rodríguez Peña

Edition: CreateSpace Independent Publishing Platform

Publication date: March 2016

ISBN: 978-1-5327-1012-4

INDEX

INTRODUCTION

This study will analyze if living with pets and the rural or urban environment enhance or decrease the diversity of the oral microbiome, and if milk and yoghurt intake can alter this microbiome. Furthermore, regional differences in Spanish oral microbiome will be studied.

METHODS

As described by Ghannoum et al. (2010), oral samples were collected at least 1 h after a meal to avoid contamination of samples with extraneous components and standardize the possible impact of variation in salivary flow rates. Study participants rinsed their mouth (swish/gargle) with 15 mL sterile phosphate buffered saline (PBS) for 1 min, and expectorated the contents of the mouth into a 50 mL centrifuge tube. The collected samples were centrifuged at 4000 rpm for 20 min at 4 ºC to separate the cells (pellet) from extracellular soluble components (supernatant). The cell pellet was used for DNA extraction, followed by PCR analysis. The 16S rRNA region from DNA sample extracts was amplified in triplicate to detect the presence of various bacteria. Bacterial PCR products were separated on the SCE 9610 capilary DNA sequencer (Spectrumedix LLC, State College, PA) using GenoSpectrum

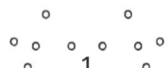

software to convert fluorescent output into electropherograms. Relative peak abundance of bacterial amplicons was calculated by dividing individual peak heights by the total peak heights in a given electropherogram using a custom PERL script. Mean normalized abundance for each amplicon was calculated from the three PCR replicates of each sample, excluding means below 1%. Normalized abundance of each peak in the electropherogram was calculated with respect to the total peak area, since it is not possible to calculate absolute abundances with LH-PCR.

In order to demonstrate whether the hypothesis is correct, the "rural vs. urban", living with pets and intake of milk and yogurt data were analyzed. The region "Basque Country, Navarre and Castile and Leon" was selected due to be the largest region with the most diverse ecosystem in Spain in order to prevent regional variables affecting this study. Subsequently the 25 most abundant bacteria in the mouth were considered for this study, considering as majority if their average were greater than or equal to 500 and as minority if their average exceeded 150. Finally, the following clusters were designed in order to do comparisons:

Pets: no pets vs dog / bird / turtle

Milk: little (<1 time per day) vs lot (> = 1 time per day)

Yoghurt: monthly (<1 time per week) vs. weekly (<1 time per day) vs daily (> = 1 time per day)

Rural vs urban: city vs rural (village + outskirts) because the outskirts also have similar levels of bacteria than the village.

Region: Basque Country vs Castile and Leon vs Navarre

RESULTS

1. Pets

The bacteria which showed significant variation are shown in these graphs:

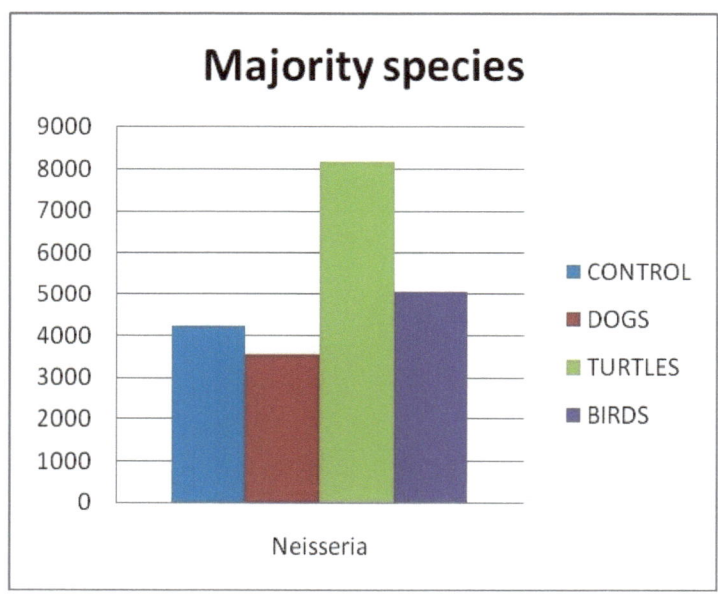

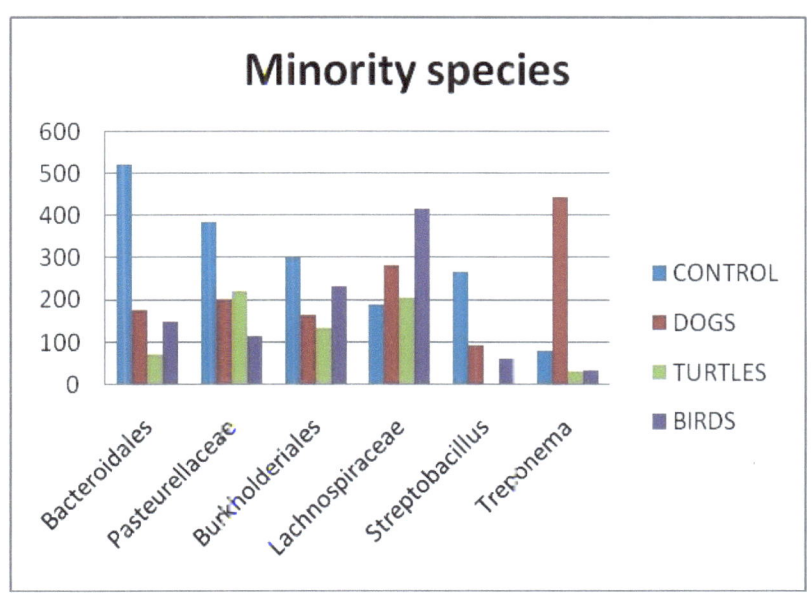

Minority species

People with dogs have a significant increase in *Treponema*, consistent with the presence of *Treponema* in the digestive tract of dogs (Weber and Schramm, 1989) and in dental plaque of dogs (Riviere et al, 1996), and the presence of the same species in humans (Riviere et al, 1996; Valdez et al, 2000).

People with turtles have a significant increase in *Neisseria*. Although the presence of *Neisseria* in iguanas is further documented (Plowman et al, 1987), th s article referred that it has also been found in turtles (Mayer and Frank, 1974).

People with birds have a significant increase in *Lachnospiracea*, which was identified in the craw of hoatzin

(Godoy-Vitorino et al, 2009) and in the intestines of chickens (Torok et al, 2011).

These increases were accompanied by decreases, perhaps by competition, in Bacteroidales, *Pasteurellaceae*, Burkholderiales and *Streptobacillus*.

2. Milk

The bacteria that showed significant variation are represented in this graph:

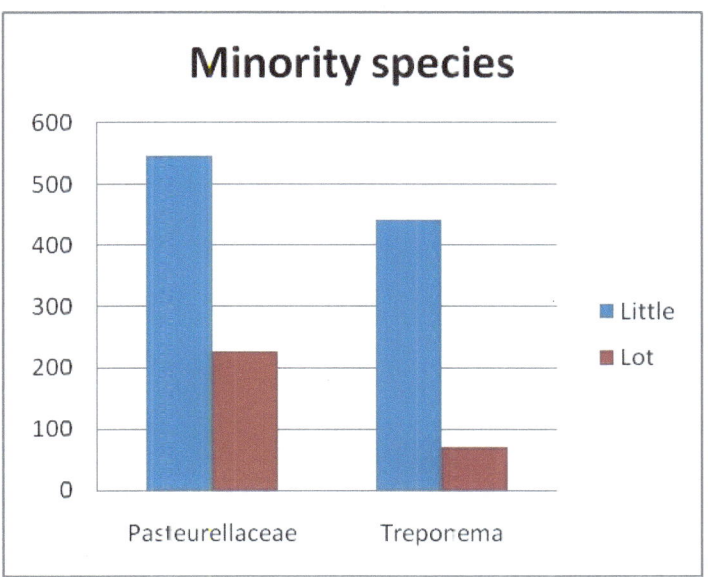

Regular intake of milk significantly reduces *Pasteurellaceae* and *Treponema*. Both species are considered late colonizers who need coaggregate with early colonizers (Kolenbrander et al, 1993). It is described coaggregation inhibition by lactose for Pasteurellaceae *Actinobacillus actinomycetemcomitans* (Rosen et al, 2003) and *Treponema* (Kolenbrander et al, 1995).

3. Yoghurt

The bacteria that showed significant variation are shown in these graphs:

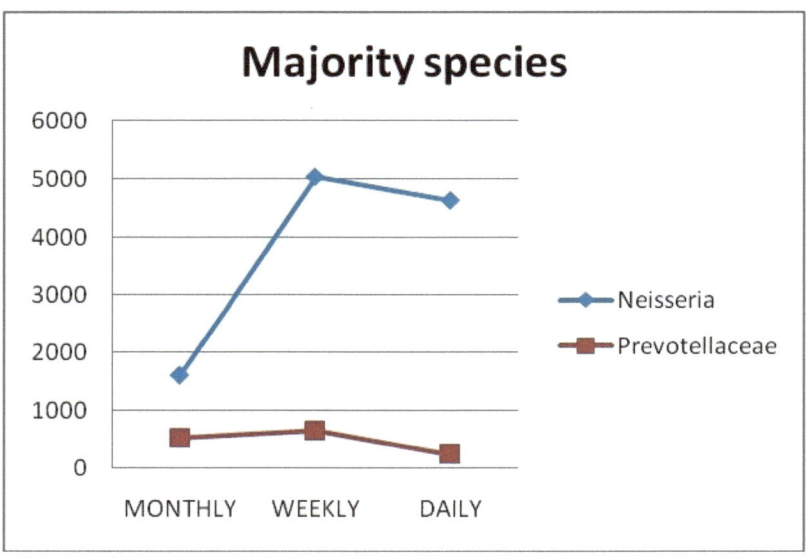

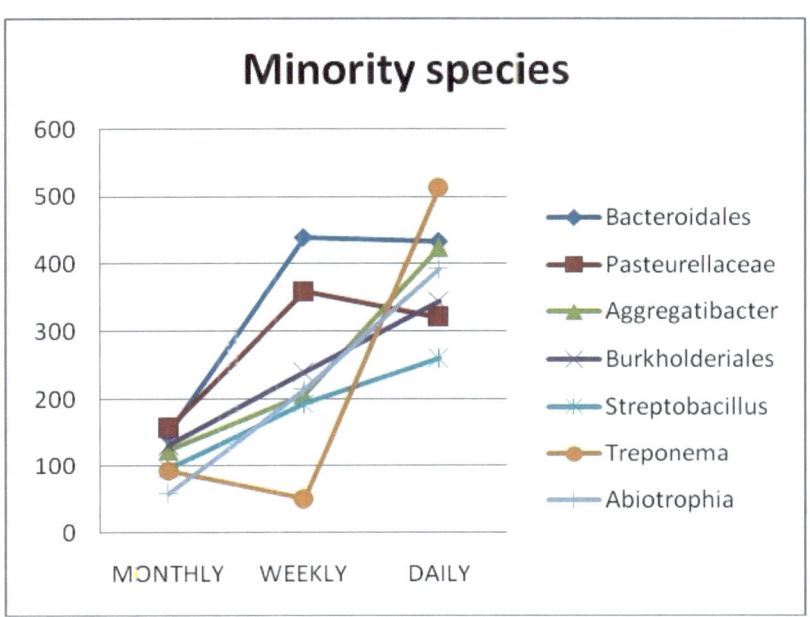

There is a significant increase in:

- *Neisseria*: it is reported that grows with lactate (Leighton et al, 2001; Smith et al, 2007)

- Bacteroidales: it grows with lactate (Brune, 2011; Tang and Edwards, 2013)

- Pasteurellaceae as *Aggregatibacter*: it metabolizes lactate (Brown and Whiteley, 2007)

- Burkholderiales: lactate is its most efficient substrate (Borole et al, 2011)

- *Streptobacillus*: *Lactobacillus* symbiont (Grigoroff, 1905)

- *Treponema*: it consumes lactate (Austin and Cox, 1986)
- *Abiotrophia*: *Treponema* symbiont (Fig 5 of Bik et al., 2010)

And there is a significant decrease in *Prevotellaceae* which can be explained by an increase in competition with Bacteroides (Koeth et al, 2013) and / or a more acidic pH which inhibits its growth (Downes and Wade, 2011).

4. Rural vs urban

The bacteria that showed significant variation are shown in these graphs:

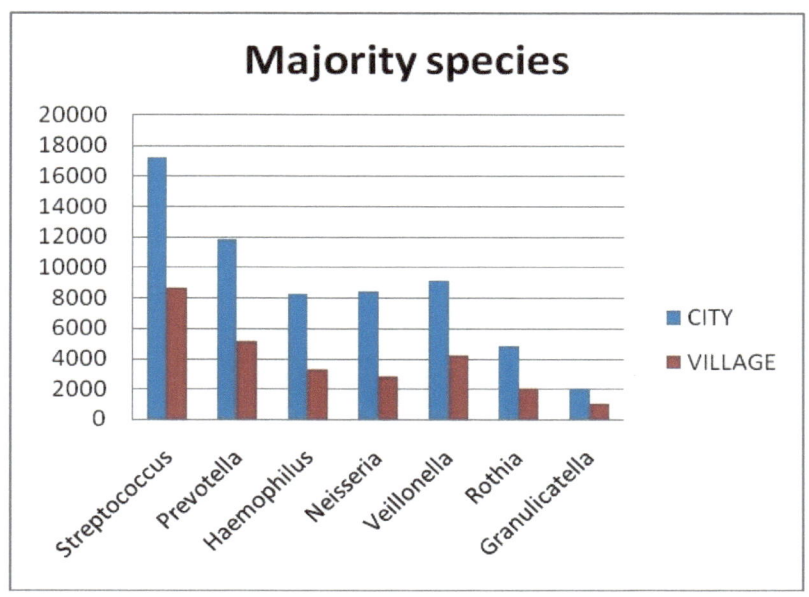

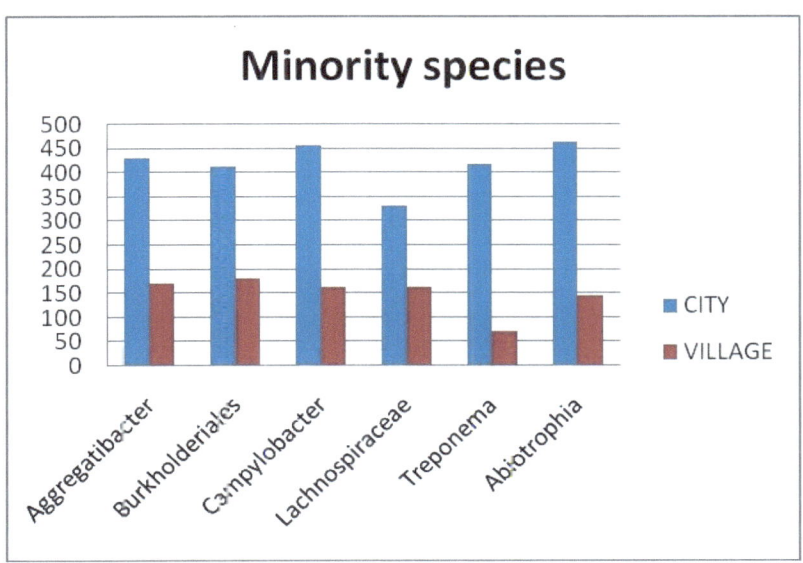

In the cities there is a general increase in many species, which agrees with Lipman (1911) who described that the cities have more dust which provides an organic substrate to bacteria, as there are fewer trees to filter out suspended particles, together with a higher concentration of people and dogs which increase their bacteria (Bowers et al, 2011) as *Streptococcus* and *Treponema* respectively, among others. It has also been detected species from bird faeces as *Lachnospiraceae* (Ryan et al, 2011).

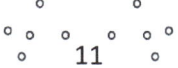

5. Region

The following graphs show the 25 studied bacteria by region whose number corresponds to the order of abundance (indicated in the table below):

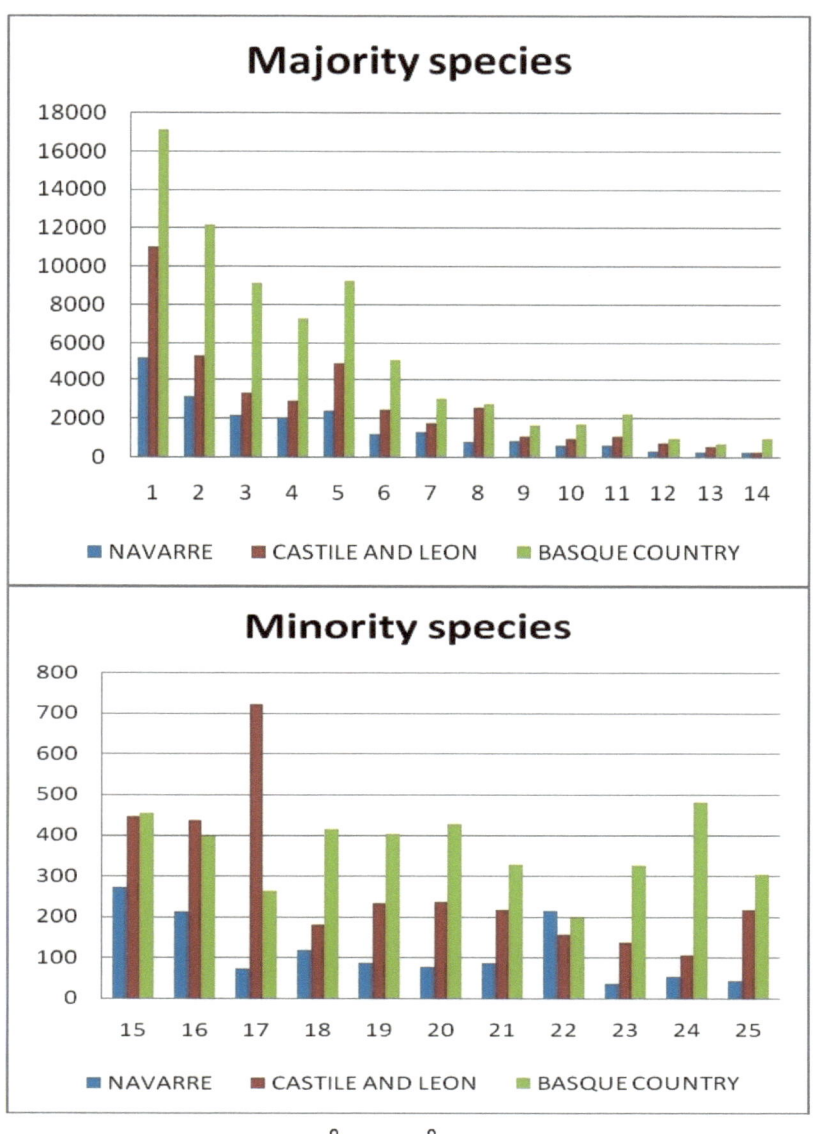

N.	Bacterium
1	*Streptococcus*
2	*Prevotella*
3	*Haemophilus*
4	*Neisseria*
5	*Veillonella*
6	*Rothia*
7	*Porphyromonas*
8	*Actinomyces*
9	*Fusobacterium*
10	*Gemella*
11	*Granulicatella*
12	*Leptotrichia*
13	Flavobacteriaceae
14	Prevotellaceae
15	Bacteroidales
16	*Capnocytophaga*
17	Pasteurellaceae
18	*Aggregatibacter*
19	Burkholderiales
20	*Campylobacter*
21	Lachnospiraceae
22	*Streptobacillus*
23	*Treponema*
24	*Abiotrophia*
25	*Megasphaera*

In Basque Country there are higher levels of bacteria probably because it is the region with the most populous city (Bilbao) besides having a wetter climate with moderate temperatures. Exceptions include the *Streptobacillus* (22) which has no regional variation, *Pasteurellaceae* (17) bacteroidal which is higher in Castile and Leon perhaps by a greater intake of cheese, and also

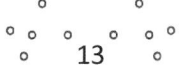

in Castile and Leon reach the same levels as in Basque Country other Bacteroidales (15), *Actinomyces* (8) and Flavobacteriaceaes (13) as *Capnocytophaga* (16). Levels of *Actinomyces* may be due to have the most fertile soils, the levels of other Bacteroidales may be due to a high intake of dairy, fish and meat (Moreiras et al, 1995) and these in turn can provide high levels of CO_2 which, together with a protein-rich diet, sustain high levels of *Capnocytophaga* (Mayrand, 1985; Sundqvist, 1992).

DISCUSION

Pets increase the levels of certain types of bacteria as *Treponema* in the case of dogs, *Neisseria* in the case of turtles and *Lachnospiraceae* in the case of birds. Contrary to expectations, milk intake hardly changed oral microbiota except a decrease in *Pasteurellaceae* and *Treponema*, however yoghurt intake increased levels of many species such as *Neisseria*, Bacteroidales, Pasteurellaceae as *Aggregatibacter*, Burkholderiales, *Streptobacillus*, *Treponema* and *Abiotrophia*. Surprisingly, a higher level of bacteria was also observed in the urban people than in the rural ones, due to a higher concentration of animals which increase their associated species such as *Streptococcus* from man, *Treponema* from dogs and *Lachnospiraceae* from birds, among other. Finally, a higher level of bacteria was found in the people from Basque Country, most likely because it has the largest city in the 3 regions, Bilbao, combined with a wetter and milder climate than in the other two regions.

REFERENCES

Austin FE and Cox CD (1986). Lactate oxidation by *Treponema pallidum. Current Microbiology 13: 123-128*

Bik EM, Long CD, Armitage GC, Loomer P, Emerson J, Mongodin EF, Nelson KE, Gill SR, Fraser-Liggett CM and Relman DA (2010). Bacterial diversity in the oral cavity of 10 healthy individuals. *ISME J. 4: 962-974*

Borole AP, Hamilton CY and Vishnivetskaya TA (2011). Enhancement in current density and energy conversion efficiency of 3-dimensional MFC anodes using pre-enriched consortium and continuous supply of electron donors. *Bioresour Technol. 102: 5098-5104*

Bowers RM, Sullivan AP, Costello EK, Collett JL, Knight R and Fierer N (2011). Sources of bacteria in outdoor air across cities in the midwestern United States. *Appl Environ Microbiol. 77: 6350-6356*

Brown SA and Whiteley M (2007). A novel exclusion mechanism for carbon resource partitioning in *Aggregatibacter actinomycetemcomitans. J Bacteriol. 189: 6407-6414*

Brune A (2011). "Microbial Symbioses in the Digestive Tract of Lower Termites" en *Beneficial Microorganisms in Multicellular Life Forms. p. 13*

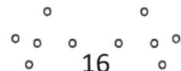

Downes J and Wade WG (2011). *Prevotella fusca* sp. nov. and *Prevotella scopos* sp. nov., isolated from the human oral cavity. *Int J Syst Evol Microbiol. 61: 854-858*

Ghannoum MA, Jurevic RJ, Mukherjee PK, Cui F, Sikaroodi M, Naqvi A and Gillevet PM (2010). Characterization of the oral fungal microbiome (mycobiome) in healthy individuals. *PLoS Pathog. 6: e1000713*

Godoy-Vitorino F, Goldfarb KC, Brodie EL, Garcia-Amado MA, Michelangeli F and Domínguez-Bello MG (2010). Developmental microbial ecology of the crop of the folivorous hoatzin. *ISME J. 4: 611-620*

Grigoroff S (1905). Étude sur un lait fermente comestible. Le "Kisselo-mleko" de Bulgarie. *Revue medical de la Suisse Romande 25: 714-720*

Koeth RA, Wang Z, Levison BS, Buffa JA, Org E, Sheehy BT, Britt EB, Fu X, Wu AND, Li L, Smith JD, DiDonato JA, Chen J, Li H, Wu GD, Lewis JD, Warrier M, Brown JM, Krauss RM, Tang WH, Bushman FD, Lusis AJ, Hazen SL (2013). Intestinal microbiota metabolism of L-carnitine, a nutrient in red meat, promotes atherosclerosis. *Nat Med. 19: 576-585*

Kolenbrander PE, Ganeshkumar N, Cassels FJ and Hughes CV (1993). Coaggregation: specific adherence among human oral plaque bacteria. *FASEB J. 7: 406-413*

Kolenbrander PE, Parrish KD, Andersen RN and Greenberg EP (1995). Intergeneric coaggregation of oral *Treponema spp.* with *Fusobacterium spp.* and intrageneric coaggregation among *Fusobacterium spp. Infect. Immun. December 63: 4584-4588*

Leighton MP, Kelly DJ, Williamson MP and Shaw JG (2001). An NMR and enzyme study of the carbon metabolism of *Neisseria meningitidis. Microbiology. 147: 1473-1482*

Lipman JG (1911). Bacteria in Relation to Country Life. *pp. 50-51*

Mayer H and Frank W (1974). Bakteriologische Untersuchungen bei Reptilien und Amphibien. *Zentralblatt für Bakteriologie, Parasitenkunde, Infektionskrankheiten und Hygiene. Erste Abteilung 229A: 470-481*

Mayrand D. (1985) "Virulence promotion by mixed bacterial infections" en *The Pathogenesis of Bacterial Infections (Bayer Symposium) pp. 281-291*

Moreiras O, Carbajal A, Campo M and Varela G (1995). Estudio nacional de nutrición y alimentación 1991 (ENNA-3). *Tomo I. Instituto Nacional de Estadística. 352pp.* Link in Spanish: https://www.ucm.es/data/cont/docs/458-2013-09-15-Moreiras-col-1995-ENNA-1-2-3.pdf

Newton RJ, VandeWalle JL, Borchardt MA, Gorelick MH and McLellan SL (2011). *Lachnospiraceae* and Bacteroidales Alternative Fecal Indicators Reveal Chronic Human Sewage Contamination in an Urban Harbor. *Appl Environ Microbiol. 77: 6972–6981*

Plowman CA, Montali RJ, Phillips LG, Schlater LK and Lowenstine LJ. Septicemia and chronic abscesses in iguanas (*Cyclura cornuta* and *Iguana iguana*) associated with a *Neisseria* species. *Journal of Zoo Animal Medicine 18: 86-93*

Riviere GR, Thompson AJ, Brannan RD, McCoy DE and Simonson LG (1996). Detection of pathogen-related oral spirochetes, *Treponema denticola*, and *Treponema socranskii* in dental plaque from dogs. *J Vet Dent. 13: 135-138*

Rosen G, Nisimov I, Helcer M and Sela MN (2003). *Actinobacillus actinomycetemcomitans* Serotype b Lipopolysaccharide Mediates Coaggregation with *Fusobacterium nucleatum. Infect. Immun. 71: 3652-3656*

Smith H, Tang CM and Exley RM (2007). Effect of Host Lactate on Gonococci and Meningococci: New Concepts on the Role of Metabolites in Pathogenicity. *Infect. Immun. 75: 4190-4198*

Sundqvist G (1992). Ecology of the root canal flora. *J Endod. 18: 427-430*

Tang S and Edwards EA (2013). Complete Genome Sequence of Bacteroidales Strain CF from a Chloroform-Dechlorinating Enrichment Culture. *Genome Announc. 1. pii: e01066-13*

Torok VA, Allison GE, Percy NJ, Ophel-Keller K and Hughes RJ (2011). Influence of antimicrobial feed additives on broiler commensal posthatch gut microbiota development and performance. *Appl Environ Microbiol. 77: 3380-3390*

Valdez M, Haines R, Riviere KH, Riviere GR and Thomas DD (2000). Isolation of oral spirochetes from dogs and cats and provisional identification using polymerase chain reaction (PCR) analysis specific for human plaque *Treponema spp. J Vet Dent. 17: 23-25*

Weber A and Schramm R (1989). The occurrence of *Treponema* in fecal samples from dogs and cats with and without intestinal diseases. *Berl Munch Tierarztl Wochenschr. 102: 73-77*

www.ingramcontent.com/pod-product-compliance
Lightning Source LLC
Chambersburg PA
CBHW041620180526
45159CB00002BC/953